In Associa

MIG WELDING:
The Plain & Simple Guide

Written and illustrated
by Chris Graham

*To Rachel, with thanks
for everything*

ISBN: 0 9525039 0 5

Published by The Busby Press © 1995
PO Box 213, Crawley, W. Sussex RH10 4YL

Introduction

MIG welding, like panel beating and paint spraying, is a skill which must be mastered if you're serious about vehicle repair or restoration. Being able to guarantee the quality of your results, time after time, is the key to success. Weld strength is important at the best of times but, becomes absolutely essential when attempting structural joints.

This booklet is intended to provide a straightforward and practical guide to the basics of MIG welding. Everything from safety and good technique to common faults and maintenance is covered in a down-to-earth style designed to complement the instructions supplied with your own machine.

The information which follows will give you the understanding and confidence needed to get started with your 'MIG'. The versatility and effectiveness of these machines is unquestionable but, it may surprise you to discover that making one work properly is not as hard as you might think. The ability to use a MIG welder effectively is one that can save you a lot of money as well as providing immense personal satisfaction.

So read on, take heart and set about developing a really useful skill that will stand you in good stead for years!

Chris Graham

Acknowledgments

Sincere thanks to Paul Ippaso at SIP (Industrial Products) Ltd and also Tony Fairweather, Matt Catton and Martyn Knowles who all provided valuable assistance with the production of this guide.

Contents

1. What is MIG?

First developed in the USA during the 1940s, MIG welding revolutionised the business of metal fabrication. It provided a more 'user friendly' alternative to traditional gas welding and knocked the brutishness of arc welding for six! Its relative simplicity made it a universal favourite, particularly with the motor industry, but this is not to say that mastering the technique is easy. The ability to use a MIG set to its full is an enviable skill indeed, and one which requires considerable dexterity, technical understanding, patience and practice.

Permanent fusion

As with all other forms of welding, the MIG process involves the harnessing of intense heat to melt the metal being joined so that it becomes permanently fused together. A trigger-operated handset or torch produces an electric arc between a wire electrode at its tip and the metal being welded. This only happens if the workpiece is effectively earthed. Without such a connection an electric circuit cannot be completed, no current will flow, no arc will form and so no weld will be possible.

A modern MIG is a sophisticated yet very usable tool. The technique needed to work one is exacting, requiring practice, patience and lots of perseverance!

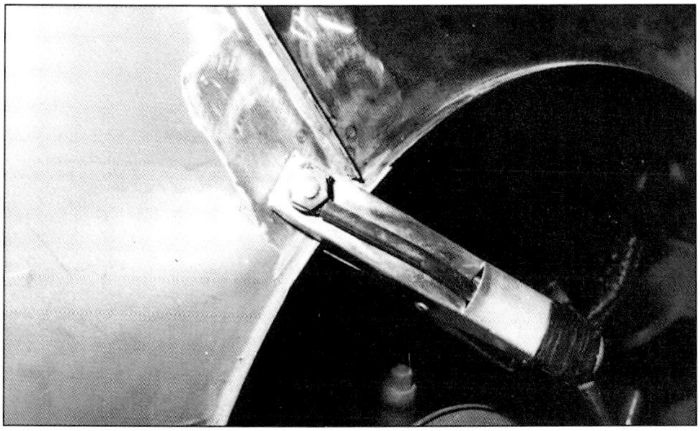

A good earth contact is essential if the arc is to 'jump'. The heavily-sprung earth lead clamp helps to ensure this assuming, of course, that the metal is clean and free from paint.

As well as being responsible for the electric arc, the welding torch also has two other very important functions. Firstly, it provides a continuous source of metal filler wire which is melted into the joint as it forms and, secondly, it supplies a canopy of shielding gas around the arc to protect the weld as it's made.

The inclusion of filler wire is vital for the ultimate strength of the weld and the addition of extra metal in this way is the key to the entire MIG process – without it complete fusion would not occur. The type of filler wire used is determined by the type of metal being welded and we'll look at the practical implications of this later.

The presence of the shielding gas is fundamental to the process as well. You see, at the instant a weld is formed, while the metal is molten, it's very vulnerable to oxidation and airborne pollution which can severely weaken it. Surrounding the weld with a very unreactive gas, known as an inert gas, provides a very neat solution to this problem. So very neat, in fact, that the whole process takes its name from it – with 'MIG' standing for 'metal inert gas'.

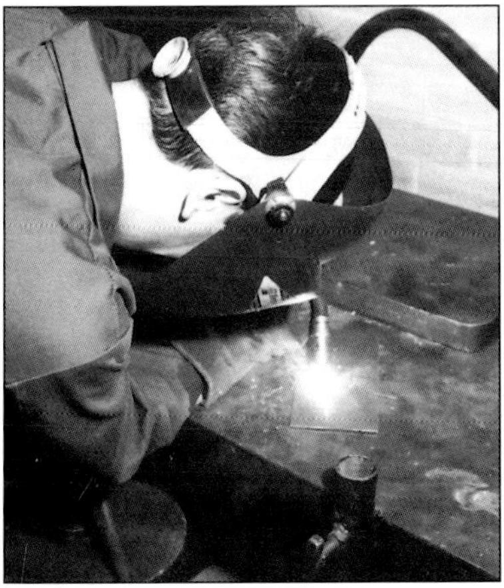

Mastering the variables – including power setting, gas supply, wire speed torch angle and movement – is the secret of quality MIG welding.

2. The machinery

Essentially, all MIG welders have the same basic construction. The only real variable is the scale of this construction and, as usual, this is linked directly to cost – the more you pay, the bigger it gets! In all cases, though, a simple steel box is used to house the electric motor (which operates the filler wire feed mechanism) together with the control circuitry, rectifier and transformer (responsible for the electric arc) and the gas solenoid valve that governs the supply of shielding gas.

Some of the more fancy models have a built-in cooling fan (often denoted by the word 'turbo' in their name) to regulate the temperature of components such as the rectifier and transformer. A MIG welder used continuously generates a lot of heat within its casing and this can reach dangerous levels if left unchecked. Consequently, manufacturers build in safety cut-out devices that are triggered once a pre-set temperature is reached. This is all well and good from a safety point of view but, in practical terms it can be a little tiresome for the experienced operator. Once the unit cuts out it has to be left to cool,

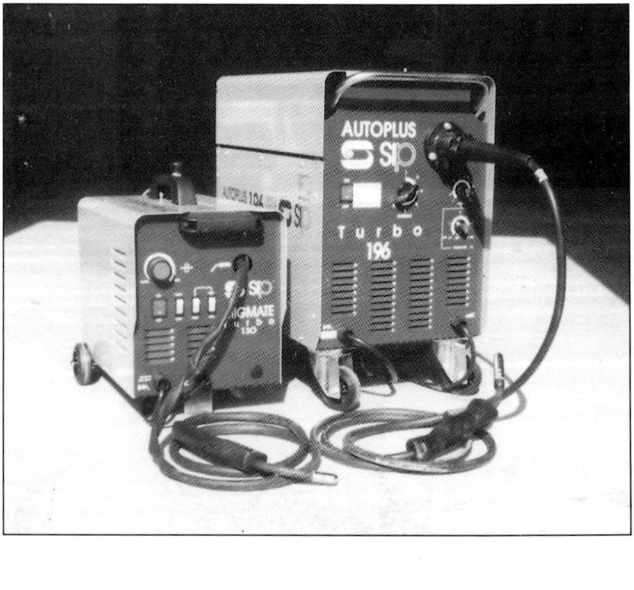

Whether your MIG is large or small, the basic principles will be the same. The big differences are controllability and, of course, price.

often for upwards of 20 minutes, before it can be used again which can become an irritating waste of time.

Fitting a cooling fan prevents this heat-related cut-out from happening which makes the welder much easier to work with for the serious user. Having said all this, however, most enthusiasts who are new to 'MIG' are unlikely to run into cut-out problems, even with the most basic gear. Frequent stops to check the weld, alter the settings, change position or simply to have a rest will act as a very effective heat control in the early days!

MIGs with a 'Turbo' badge, or 'fan sign' may seem flash but this denotes the presence of a cooling fan – a valuable feature indeed.

Keeping control

If you've bought a small MIG welder then the controls, all usually located on the front panel, will be simple. You'll find an on/off switch, a control for the wire speed and one or more voltage level selectors. In addition, you will also notice three cables issuing from said panel. The thickest of these is the one which carries the filler wire, the shielding gas and current to the torch and, although it's armoured, it should always be treated

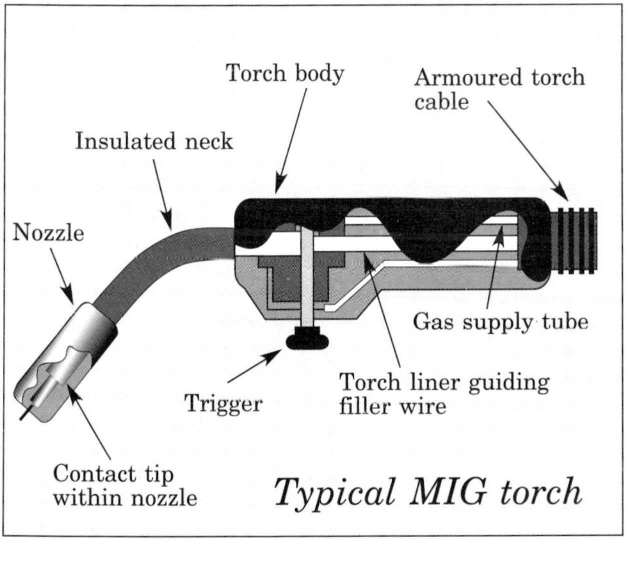

The MIG torch is a pretty simple device which requires next to no maintenance. Reliability is good and failures are rare.

Controls on the average DIY MIG are pretty basic. However, the more you pay, the more knobs you get to twiddle!

with respect. Never wind it tightly, kink it or stand on it.

The other two cables will be conventional electrical ones. The one with the spring-loaded clamp on the end is the earth lead which has to be connected (via bare metal) to the workpiece to complete the circuit. The third, probably with a three-pin plug at its end, should need no explanation here!

The control for the wire speed is normally in the form of a knurled wheel (either with or without click stops) which is rotated clockwise to increase the speed and *vice versa*. Often there will be some form of calibration printed around the wheel but, when it comes down to it, there really isn't much significance to this as good wire speed setting, as we'll see later, is borne out of experience and little else.

Other choices

The number of different voltage settings, often selected with rocker-type controls on the smaller machines, depends on the number of switches. Often, in a welder's instruction leaflet (or sometimes on the front panel itself) a guide linking particular settings with different thicknesses of metal will be provided. Never take this information as gospel, however. It should only ever be regarded as a starting point and, once again, it is experience which should eventually become your guide in this respect – there is simply no substitute!

Larger, more expensive machines usually boast, as you might imagine, a rather more complicated set of controls. Included among these are additional features such as automatic timers and switches for different function settings (for making alternative types of weld) – all part and parcel of these higher-spec., costlier units.

The wire speed control must be set correctly for effective MIG welding. Poor setting is a common error made by beginners.

Gas supply

The shielding gas used in MIG welding comes in different types, and I'll be explaining the reasons for this in a later chapter. However, irrespective of type, the gas used will always have to be supplied from a pressurised cylinder, the size of which depends on the specification of the machine you have. All 'starter' MIGs these days work from (and are often sold complete with) the small mini cylinders or canisters which are readily available from most good DIY and motorist discount stores.

A simple, brass 'on/off' gas regulator screws into the top of the cylinder (to reduce the pressure) and is linked to the welder itself by a thin, flexible plastic tube. This is usually a straightforward push fit and its job is to feed the gas into the machine where it then joins the wire on the short journey up

Mini gas canisters feature a simple brass on/off regulator. Make sure you don't damage the thread while attaching it.

The shielding gas leaves the canister through a thin but flexible plastic tube. This is normally a simple push fit into the regulator.

the reinforced cable to the torch and out through the nozzle.

These small gas canisters are ideal for getting started. Their portability and good availability make them an attractive initial proposition. However, they do have their disadvantages. Firstly they are relatively expensive and, secondly, their contents don't last particularly long. Opinions on this differ. Most manufacturers will tell you that they contain enough gas for 30 minutes of welding but practical tests sometimes tell a different story. In the hands of beginners the canisters may only last half this time due to poor technique, leaks etc.

Much also depends on the type of gas. For example, CO_2 exists inside the cylinder in liquid form which evaporates to a much greater volume than that provided by the compressed gases (argon, argon/CO_2), therefore, it will always last longer.

Cylinder size?

The ideal solution, once you're up and running, is to upgrade to a larger cylinder which will prove far more economical in the long run, assuming high usage. Being practical, though, the deciding factor in this respect must be cost. If, as is the case with most beginners, your MIG is only going to be used occasionally then such a switch is probably not worth the bother. Remember, larger cylinders can be very inconvenient to source and, to use one, you'll need a good quality gas regulator and the pipework and adaptors to go with it which adds considerably to the cost. They are, of course, refillable but only by recognised outlets such as BOC depots.

Necessary tension

Lastly, at this stage, it's worth devoting a word or two to the wire feed mechanism. As you should realise by now, this is controlled, together with the gas supply and the current, by the trigger on the torch. The supply of filler wire is central to the whole MIG process and if its tip is not kept sufficiently close to the workpiece then the arc will not form. Primarily this distance must be maintained by the operator's skill in holding the torch steadily, but the speed of the wire's movement up to the torch plays an important part too.

In most cases unclipping the removable side panel or opening the door/lid on your welder will reveal a spindle on to which the reel of filler wire is fitted, together with a pair of grooved rollers through which the wire must pass on its way to the torch. In fact, it's these rollers which provide the driving force for the wire. They are adjustable, in terms of the pressure with which they bear on one another and this adjustment is an important factor.

We'll consider, in more detail, the implications of the wire feed tension in due course but, for the time being, it's important to realise that such an adjustment exists and that it serves a valuable function. Without the correct tension on the wire, the MIG will not work properly – it's as simple as that!

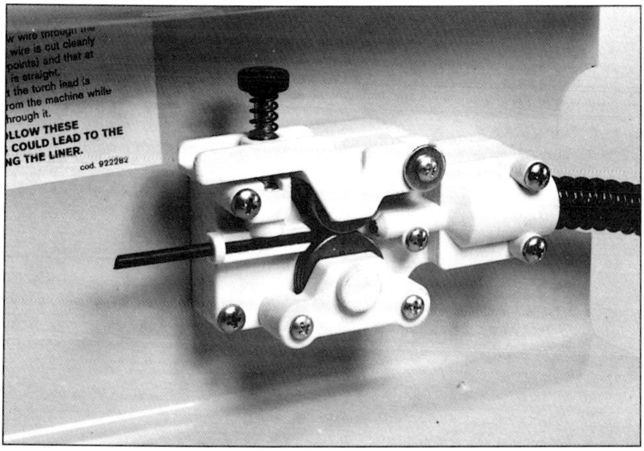

The wire tensioning mechanism on a MIG welder is not complicated but it must be used properly. Driven rollers power the wire up to the torch, the tension between them is crucial.

3. Gas, wire & volts

Success with a MIG welder is all about controlling the environment immediately around the arc. Knowing how and why to adjust the variable factors is half the battle and it's only with correct use of the settings that good, strong welds can be achieved.

The type of gas you choose is one of the more basic decisions to be made before you start. The alternatives are not that many but the differences between them can be pronounced.

The importance of gas

We've already established that the arc has to be shielded from the atmosphere to perform its function properly. Failure in this respect leads to oxidation and other contamination of the newly-formed weld which is bad news for the longevity and strength of the final joint. Surrounding the arc with a very unreactive gas (an inert gas) is the best way of guaranteeing against the destructive effects of the oxygen, nitrogen and hydrogen which are naturally present in the atmosphere.

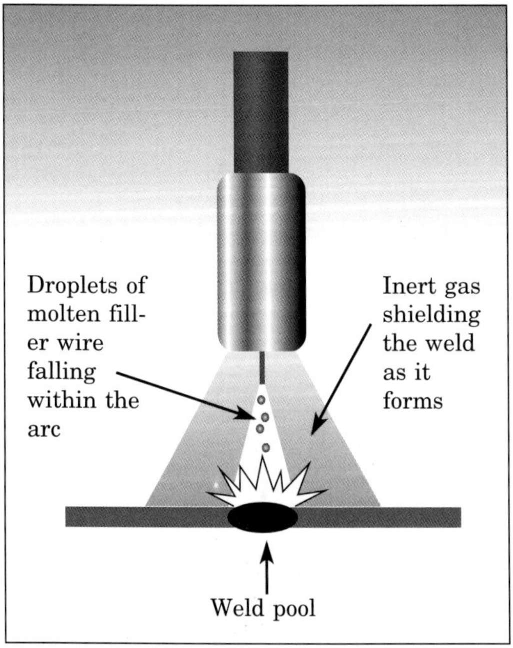

Different metals require the presence of different shielding gases if effective welds are to be made. As a general rule, when working with mild steel a user must select pure carbon dioxide or a mixture of argon and carbon dioxide (commonly

It's vital that the gas used is matched to the metal being welded. Most DIY MIGs are supplied with Carbon Dioxide these days.

95% and 5% respectively). Stainless steel welding calls for a mixture of argon and oxygen (just 1% – 5% oxygen) while the joining of aluminium demands an atmosphere of pure argon only.

The gas most commonly used at the DIY end of the MIG welding market is carbon dioxide (CO_2) and this is what you will usually find supplied with the popular 'starter' machines these days. The main reason for this is its cost – it's the cheapest of the lot! However, as we all know the cheapest isn't always the best and many experts consider that a better, more controllable weld results from using a mixture of carbon dioxide and argon. You see, CO_2 is not completely unreactive and this fact can lead to the arc getting a little frisky!

The upshot of this 'excitement' is that droplets of molten metal, instead of falling directly from the tip of the filler wire and down into the area of molten metal known as the 'weld pool', get thrown out to settle and solidify all over the surrounding metal in the form of what's called 'spatter'. While not an absolute disaster, this phenomenon can be a considerable

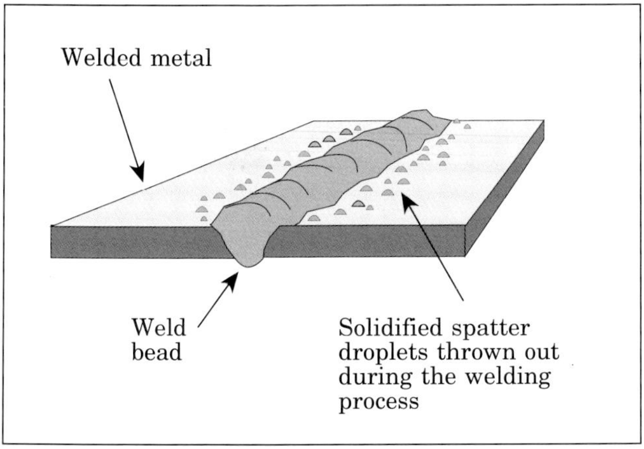

'Spatter' around a weld tends to look worse than it really is. It is a common fault and is best dealt with using an electric grinder.

nuisance because the solidified metal 'droplets' will, in many cases, have to be cleaned off with a grinder afterwards. Obviously this doesn't really matter with some jobs where appearance is not important but, when welding a new section into a wing, for example, where neatness is visually important, having to deal with spatter can be a bind.

Why a mixture?

Opting to work with a gas mixture greatly reduces the risk of spatter because the presence of the argon (a truly inert gas) really does calm things down, leading to a much more stable arc and a smoother weld. A further benefit is that the weld runs at a lower temperature than with pure CO_2 which means that the likelihood of hole-burning in thin metal sheet is considerably reduced – a distinct benefit for the beginner! Any lessening of the weld's temperature is also good news when working on thin sheet metal because lower temperatures mean less heat-related distortion.

In the same way that the shielding gas has to be matched to the metal being welded, so does the filler wire. If anything, though, it's perhaps even more important to choose the right wire because this is what's actually helping to produce a strong joint between the two edges being welded.

Dissimilar metals simply cannot be used for MIG welding. Successfully welding together two pieces of mild steel calls for

As with the gas, there are important choices to be made about the type of filler wire used. It, too, must be matched to the type of metal being welded.

the addition of mild steel filler wire. Correspondingly, welding stainless steel requires stainless steel wire, and so on. In short then, the filler wire used must match the metal being welded. It is hopeless to try to weld mild steel with aluminium wire because the two will not mix when molten and complete fusion will be impossible.

Which wire?

It is likely that most DIY-type MIG welding machines will be supplied complete with a small reel of mild steel filler wire (1kg/2lb). This locates on to the spindle inside the machine from where the wire passes through the pair of tensioning rollers and on up the guide tube towards the hand torch and, ultimately, the weld itself. Professional machines tend to make use of larger (5 – 15kg/11-33lb) reels of filler wire but you will find an increasing tendency among the smaller units nowadays for them to have the capability to accept a 5kg/11lb reel as well. It is, of course, more economical to buy your filler wire in as large a reel as possible but much obviously depends upon how frequently you will be using the machine. Bare in mind, though, that mild steel wire will rust if left in a damp garage for any length of time. Removal from the machine and storage in a dry place is the sensible approach if use is only occasional.

Don't be surprised when you discover that the mild steel filler wire on your new spool has a decidedly pinkish-brown

The larger the reel of filler wire, the more economic it is to buy. Much, of course, depends on how often you use your machine. Regular users will benefit if their machine can be fitted with a bigger reel like this.

Remember that the diameter of filler wire that you can use in your machine is governed by the width of the groove in the wire tensioning roller.

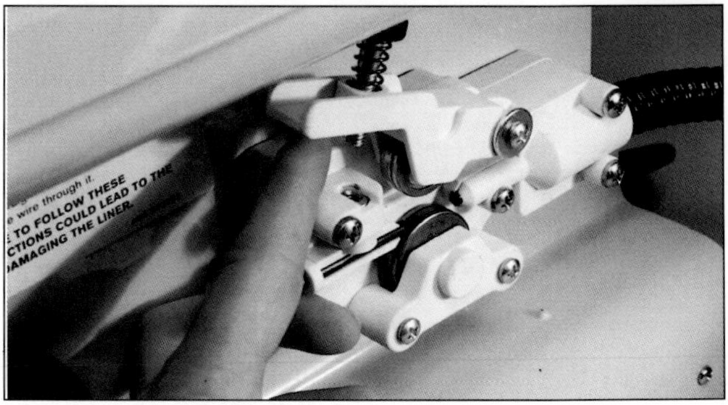

tinge, rather than being silvery as you might have expected. This is quite normal and is caused by the fact that, during manufacture, the wire is coated with a thin layer of copper as an anti-corrosion precaution.

Wire diameters

Filler wire, as well as being supplied in varying types, is also available in different diameters. The common ones are 0.6mm/0.024in, 0.8mm/0.030in and 1.0mm/0.039in and the decision about which of these to use is based broadly on the type of work being undertaken. As a general rule it's sensible to use 0.6mm/0.024in wire on thin sheet metal, 0.8mm/0.030in for thicker metal and more general-purpose welding jobs. In many cases small MIG units are nowadays supplied with 0.6mm/0.024in because the manufacturers assume that most enthusiast users will be working on car bodywork etc. The use of 1.0mm/0.039in filler wire will normally be beyond the capability of most small MIGs because they simply don't have the power plus, of course, they will need additional wire rollers with wider grooves to handle the larger diameter.

Without an electrical supply the technique of MIG welding would not be possible. If the electrical arc were not to jump between the tip of the filler wire (consumable electrode) and the earthed workpiece, then no heat would be generated.

Plugging your new welder into the mains and flicking the switch sends 240 volts or so rushing down the cable and straight into a transformer where it's smartly converted into a more manageable, low alternating voltage. From here it passes into the rectifier from whence it emerges as a DC voltage.

What is known as a choke coil then stabilises the current after which it is ready for transmission to the torch and its jump, in the form of an arc, down on to the metal subjects being welded.

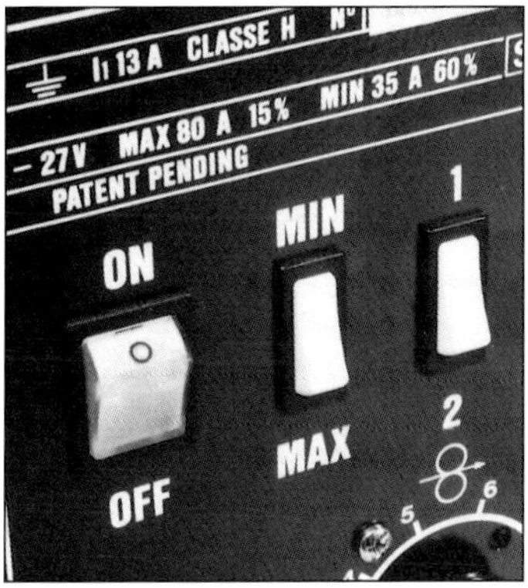

All MIGs feature voltage controls, normally in the form of rocker switches like these.

Controlling the voltage alters the severity of the arc and, therefore, has an important bearing on its power. Even the most basic units will feature a facility for selecting a number of different voltage settings, usually in the form of rocker switches. These have to be altered, in conjunction with the speed of the wire feed, to best suit the thickness of metal being welded.

Power play

Knowing which setting to select is largely a matter of experience but many suppliers do now include guidelines about this with their instructional literature. These provide a basis on which to work but should only ever be regarded as a starting point. It really is pretty well impossible to generalise about power settings because of the number of other variables (metal type and thickness, joint type, gas type, machine type, working environment, skill of the operator etc.) which must be considered. In later chapters, however, we'll look at the practical effects of variations in the voltage setting in particular instances and, in this way, you'll start to appreciate the crucial importance of making the right voltage selection for any given welding job.

Beware of setting suggestions from the MIG manufacturers. They are only intended as a guide and should never be regarded as anything more than that.

4. Be safe, not sorry!

Possessing a full appreciation of the safety aspects associated with MIG welding is essential for the preservation of body and soul! Although completely safe in experienced hands, a MIG welder is, potentially, a very dangerous piece of equipment if misused and so should always be handled with the greatest respect. Remember, you're dealing with mains electricity, intense light, pressurised gas and extreme heat so the proper safety precautions really are vitally important.

To begin with always make sure that your machine is properly wired-up before you plug it in to the mains supply. Most electrical appliances are now supplied complete with a three-pin plug attached but it's still a sensible idea to spend a minute or two making sure that this is wired correctly (assuming it isn't sealed) and has the right fuse fitted.

Dangerous business

The two by-products of the electric arc are heat and light. Both present potential dangers for the careless user. The considerable heat (3,000°C) generated by the arc rapidly spreads through the metal being welded and can cause all sorts of problems. Most obviously there is the risk of straightforward skin burns if you accidentally touch the job before it's had time to cool – a silly mistake and one that most tend to make just the once!

A knock-on and perhaps even more serious effect of 'heat spread' can be fire or explosion. If you are attempting to weld car body panels it's vital that they are free from old paint, underbody sealant, filler, sound deadening materials or anything else that might present a fire risk. It's all too easy to clean off the bits you can see but completely overlook those which you can't. With these thoughts in mind you should always have an appropriate fire extinguisher at the ready just in case. Mistakes can be very costly.

The other big heat-related danger is that of explosion, most commonly caused, in the case of vehicle restoration, by the ignition of petrol vapour. MIG welders and petrol tanks are not the best of bed fellows and yet it's surprising just how many accidents are caused by people putting them together! Always be sure exactly where the petrol tank is before you start welding and, if in any doubt at all, remove it altogether to eliminate the risk. Simply draining the tank is no guarantee of safety because the vapour that's inevitably left is just as serious a hazard as the petrol itself.

Lastly, on the subject of fires, it's worth bearing in mind that MIG welding produces sparks which can fly anything up to 10 feet away from the torch. You'd be wise to make sure that cans of petrol, tins of paint, old oily rags and all other combustible products are kept well out of reach. When you're concentrating hard behind a headshield it's easy to fail to notice the beginnings of a fire, even when it's only feet away.

Eye contact

Most of you will associate MIG welding with the strangely attractive, flickering blue light that's given off. This, of course, is produced by the electric arc and, for all its visual fascination, is not something to be gazed at. Being rich in ultra-violet, this intense light source is harmful to the skin. Hands, arms

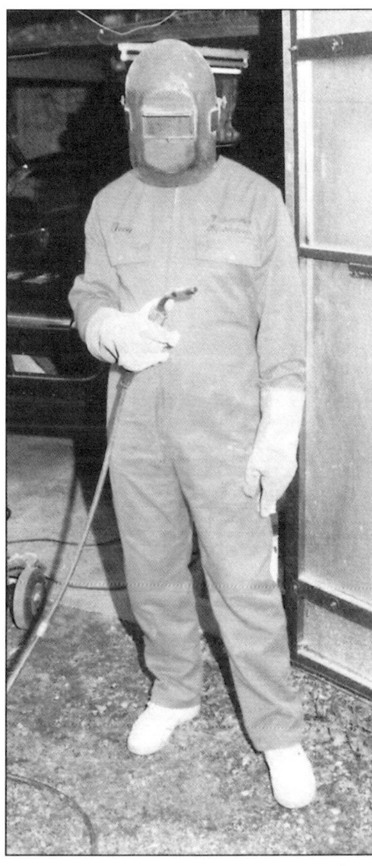

Full protective clothing is essential when using a MIG welder. Never take chances – it's just not worth the risk.

and faces will quickly get burned if exposed to this light, which acts in the same way as the sun's rays, only more ferociously.

The other big risk from the arc is to the eyes which can become badly affected if left unprotected. The intensity of the light is capable of seriously damaging the front surface of the eye, called the cornea. An exposure of just a few seconds can be enough to induce a painful and unpleasant condition known as 'arc eye' or 'flash', in which the surface of the eye becomes dried out, sensitive and very painful. In most cases these symptoms last for a few days but in severe cases the problems can be more serious.

Protecting the eyes from the arc involves the use of a face shield. These are available in two basic types but both feature a built-in darkly-coloured filter through which the user observes the weld. Filters for welding, which are usually coloured green, are calibrated on a simple numerical scale, with the darker tints getting the higher numbers. However, all filter lenses which comply to BS639 will give total protection to the eyes against IR and UV radiation so check that this important specification applies to any equipment you buy.

Professionally speaking

The most basic type of headshield is the small, hand-held variety generally supplied with most DIY-type machines. These are

Welding shields can either be hand-held (left) or worn on the head. Most small MIGs are supplied with cheap examples of the former. Serious users normally choose to work with the latter.

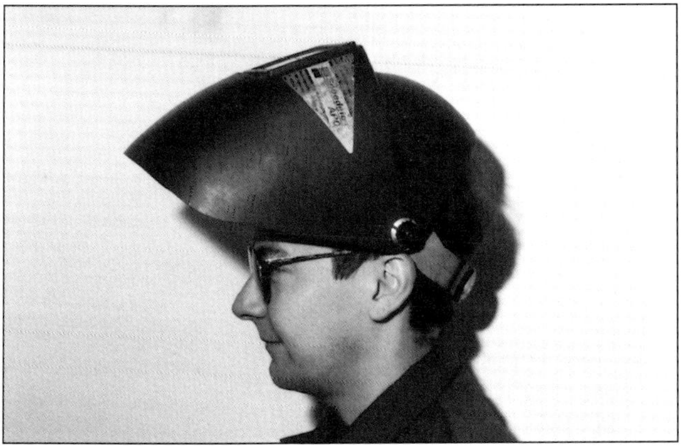

The big advantage of a hinged headshield is that it leaves both hands free to control the torch. It also provides better protection too!

all very well as far as they go but you would never catch a professional using one. For a start, their ping-pong bat-like shape means that complete facial protection cannot always be maintained. It's true that their size does vary from supplier to supplier but some are so tiny that they are hard to use effectively.

Quite apart from the dubious protective qualities of the hand-held mask, the single biggest problem with them is that they *are* hand held. Their use means that the operator only has one hand with which to hold and support the torch. Now, you may imagine that this is of little consequence but I can assure you that it makes a big difference to the ease, comfort and accuracy with which the weld can be made.

Fortunately, there's a convenient solution to the 'hand-held' problem and it's to dispatch the flimsy article smartly into the nearest bin and to buy yourself a hinged welding helmet! They are not terribly expensive but will revolutionise the way you work. Because of its size and shape, this type of head shield provides complete protection from the arc and, being hinged on an adjustable head band, leaves both hands free to take care of the torch.

The ultimate in headshield technology so far is the light-sensitive filter. Controlled from within, the filter darkens instantly the arc starts.

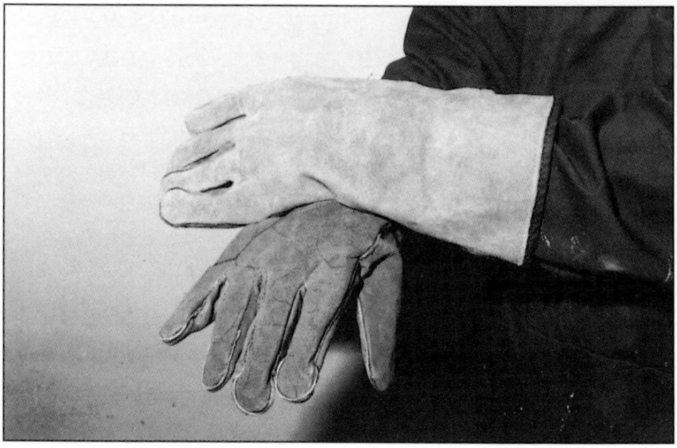

Always wear thick leather gauntlets when using a MIG welder. Make sure, also, that the sleeves of your overalls are tucked in to protect your forearms.

The ideal combination is what's known as an 'autofilter' head shield which features a light-sensitive filter lens which remains clear until the instant the arc forms and welding begins. At this point, within a fraction of a second, the filter darkens sufficiently to protect the eyes and remains so until the arc is switched off. Unfortunately, though, swish accessories like this don't come cheap although it should be noted that they are a good deal more versatile than conventional head shields. The fact that the filter lens only darkens in response to the arc light means that this type of shield can be usefully employed for other jobs where facial protection is important like grinding, chipping and scraping.

Keep it covered

What you wear while playing with your new MIG welder is also of considerable importance. Bermuda shorts and a 'U2' T-shirt really are out of the question! Remember that every spark you see leaving the arc is, in fact, a tiny droplet of molten metal. These little critters love nothing better than to tumble up your sleeve, sneak inside your shoe or drop down the front of an open-necked shirt – whichever happens, it really does hurt and is certainly best avoided.

Sensible operators wear good quality, tightly-woven cotton overalls which button right up to the neck. On their hands they will sport a pair of leather gauntlets extending up to the elbows and, on top, a light hat will be worn under the head shield to protect the top of the head.

It is obviously essential that all protective clothing is non-flammable. Garments made out of nylon and other synthetic materials are hopeless because they melt, stick and burn like mad. Also make sure that overalls are clean – patches of soaked-in oil are not recommended.

The last points to make on the subject of welding safety relate to the environment in which you are working. The fumes produced by the MIG process can be toxic and so it's important that your workshop or garage is well ventilated. This is especially so if you are working in a confined part of a car's interior, such as a front footwell or in the boot. Avoid, however, setting up a fan to blow on the welding area as this will more than likely upset the gas shield and ruin the quality of the weld.

Think of others

If you have inquisitive young children or pets then make sure that they are well out of the way before you start. Young, staring eyes are all too easily damaged so be careful. Also, if you're working outside spare a thought for the neighbours. People unfamiliar with the technique are apt to get decidedly uppity when confronted with crackling arcs, buzzing wire feeds and showers of sparks!

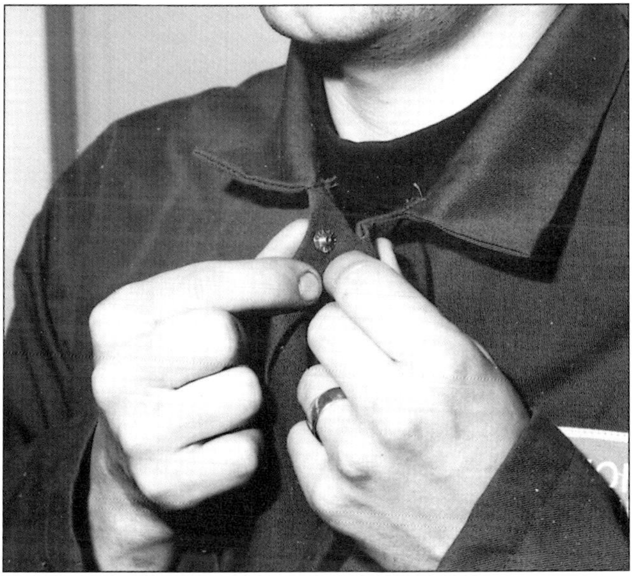

Overalls for welding should be made of tightly-woven cotton and must be clean. Always fasten them up to the neck for your own protection.

5. Getting started

A MIG welder, unlike an electric kettle or a hair dryer, isn't really the type of machine that you can use 'straight out of the box', despite what the enthusiastic salesman would have you believe! The exacting nature of the work dictates that care and attention to detail is needed when readying a MIG unit for action. There are some important things that must be done properly if the machine is to function well and safely, so you'd be wise to take your time at this stage. Race through the setting-up procedure and the machine will never work properly.

Basic preparation

Your new MIG is likely to come supplied with one gas canister (together with a screw-in regulator) and a spool of wire, both of which will have to be attached to the machine before you can do anything. Dealing with the gas bottle should be pretty elementary; it usually fits on to the back of the machine and is secured with a combination of clips, brackets, straps or chains, depending on the design.

With the canister securely in place, the plastic dust cap should be removed from the valve so that the gas regulator can be carefully screwed on without damaging the fine thread. Remember that it's important to check that the regulator control is switched off before fitting it otherwise some of the precious shielding gas will be wasted. As already mentioned, the gas itself is piped into the

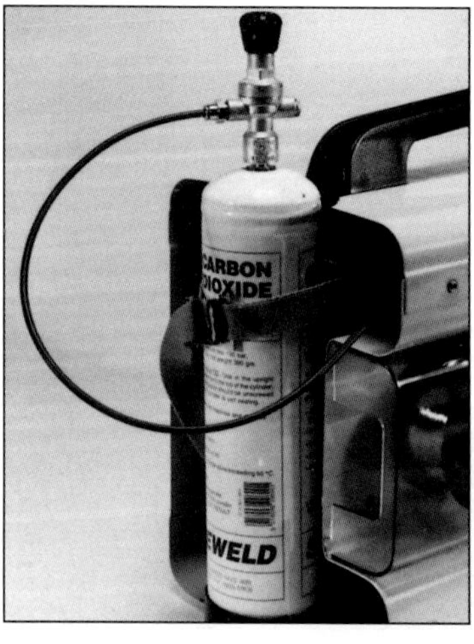

Mini gas canisters must be firmly secured on to the MIG. Most machines have simple fixing mechanisms like this.

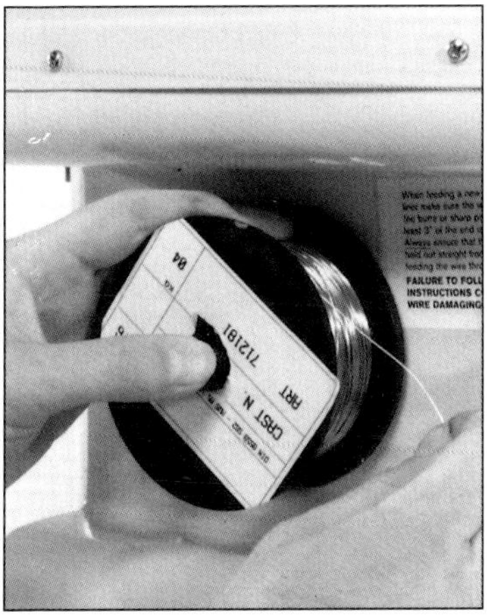

When fitting the spool of filler wire be sure that the wire leaves the reel from the top, as here.

welder via a small, flexible plastic tube which is normally a simple push fit into the regulator.

Fitting the wire spool is a little more involved but the procedure should be fairly well detailed in most instruction booklets. Basically the spool is mounted on a spindle found behind the removable panel or hinged door in the outer casing of your machine. A spring-loaded nut (or similar mechanism) is used to secure the spool to the spindle. This should be set tightly enough to prevent the spool from spinning freely but, at the same time, allowing it to rotate smoothly as the wire is drawn off. A free-wheeling spool will lead to slackness in the wire which goes on to cause all sorts of nasty problems.

Just ahead of where the spool is mounted you will see the wire tensioning assembly with its twin rollers and screw-type adjustment. This is a very important part of the whole machine because it controls the movement of the filler wire up to the torch. One of the rollers, normally the lower one, is dri-

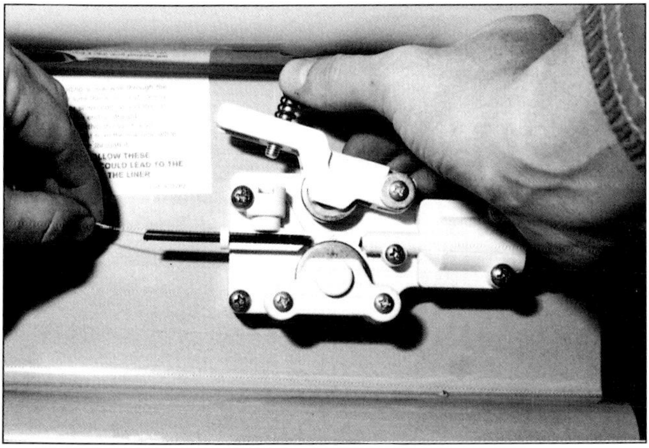

Pass the filler wire through the first section of guide tube, then between the two tensioning rollers and on into the remainder of the guide tube.

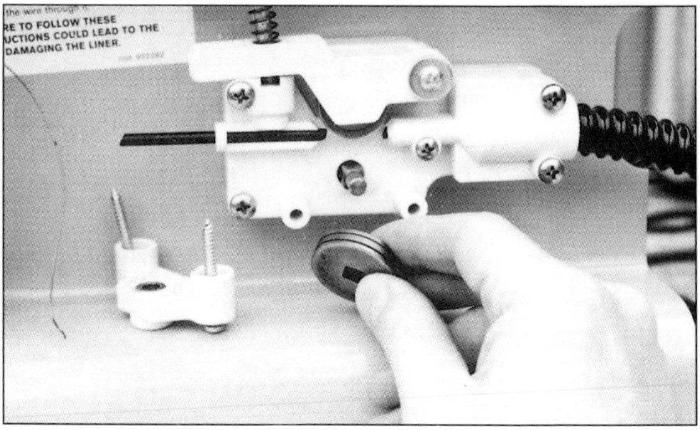

Before clamping the rollers make sure that they are the right ones for the diameter of wire being used.

ven by a motor, the variable speed control for which will be on the front panel of the welder. This bottom roller will also be grooved (often with two channels) to provide a guide for the wire as it passes. Where two grooves are evident the roller will be reversible allowing for the use of an alternative wire size. Make sure the groove selected matches the diameter of the wire being fed through the machine – a simple but common mistake to make.

The tension mounts!

The top roller is designed so that it can be clamped down, sandwiching the wire and ensuring a good grip. Altering the

At this stage you should also check that the contact tip under the torch nozzle is the right size for the wire.

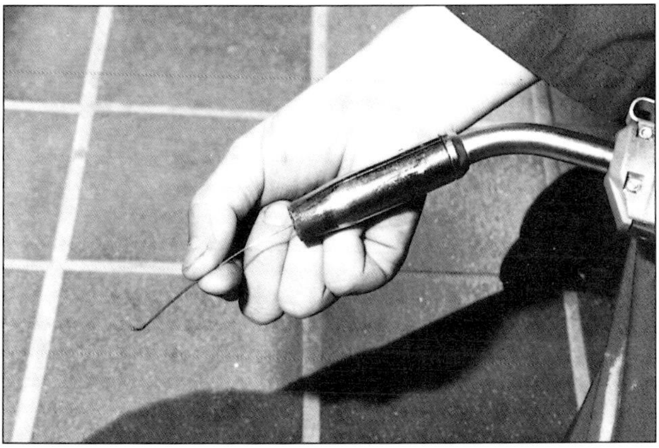

Setting the wire tension correctly is important. Light finger and thumb pressure should be all that's required to induce wire slippage. Any greater effort means that the wire tension is too great.

clamping action changes the pressure on the rollers to govern the force with which the wire is transported up to the torch.

Before you start threading the wire through, snip off the tip with wire cutters to produce a 'clean' end so that there is no risk of it snagging *en route*. Then pass the tip through the inlet guide tube, undo and separate the rollers, settle the wire into the groove in the bottom roller and feed it on into the continuation of the guide tube which will conduct it the rest of the way up to the torch. Next clamp down the top roller (not tightly) and switch on the power supply to the MIG.

Unscrew or pull off the nozzle from the end of the torch, then undo and remove the contact tip with pliers. Depress the trigger on the torch (gas still switched off) at which point the motor will start and the rollers will begin to feed the wire up to the torch. When it emerges, clip the tip off again and then test the fit of the contact tip over the wire. It should be stamped with its size (to match the wire) but you can never be sure until you've tried it. If all is well, tightly secure the contact tip and replace the nozzle.

It only then remains for the tension of the wire to be set. Getting this right is important so take care and never assume that it's OK until you've checked it. The simplest practical test of wire tension is to grip the wire between finger and thumb as it leaves the torch and operate the trigger so the feed motor starts. Set correctly, light finger and thumb pressure should be sufficient to induce slippage and stop the wire (refer to the photograph above).

6. Good technique

Adopting the right technique is one of the fundamentals for successful MIG welding. Without it any hope of a strong, durable joint is just pie in the sky – it really is that crucial!

To be a good welder calls for a certain level of manual dexterity, reasonable co-ordination and a steady hand. I think that it's fair to say that much of the basic technique can be learned by the beginner but, if he or she does not possess the three 'natural' qualities already mentioned then, ultimate ability is bound to be limited. Some enthusiasts never manage to master the 'art', however hard they try, but most do achieve an acceptable level of performance.

Ready, steady...

The first thing to remember is that for the best results you must be comfortable. A comfortable welder is a relaxed welder, and a relaxed welder produces a much better result. Stance is important, as is balance. Stand in a comfortable, well-balanced manner with your feet slightly spread to provide a stable foundation. Regard your whole body as an elaborate support mechanism for the welding torch, because that's what it must be.

Always adopt a relaxed stance when using a MIG welder. You'll work much better and more accurately if you prevent yourself from tensing up.

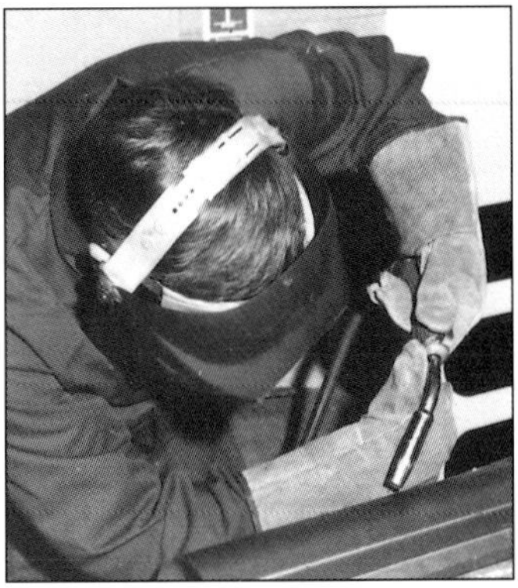

Buy yourself a hinged head-shield so that you can have both hands free to support the torch. This really does make life so much easier for the beginner.

The movement, angle and stability of the torch are the all-important factors behind the MIG welding process.

Equipping yourself with a full-face head shield allows both hands to work with the torch which makes life a lot easier. Most professional welders will use their 'spare' hand as a support for the torch as it progresses along the joint. Having to struggle with a hand-held face shield prevents such 'luxury' and makes the job a lot more difficult, not to mention tiring.

Heading for success

The position of the head is another very important factor. It's essential to have an unimpaired view of the weld as it's being made so you must get your head down so that the arc can be observed from the side or front (whichever you prefer). Don't worry about putting your head too close to the arc because,

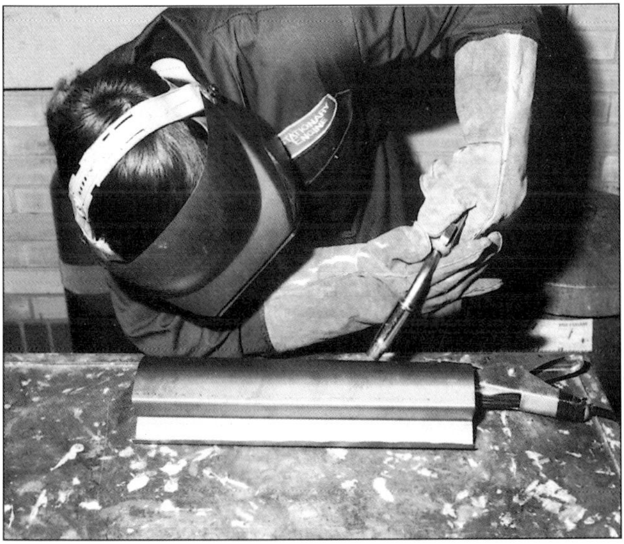

Getting your head in the right position is very important if your welding is to be a success. You should position it like this, ahead of the torch and close to the arc for the best possible view.

Don't do it like this! Holding your head directly above the torch will obscure your view of the arc and significantly impair your control over its progress.

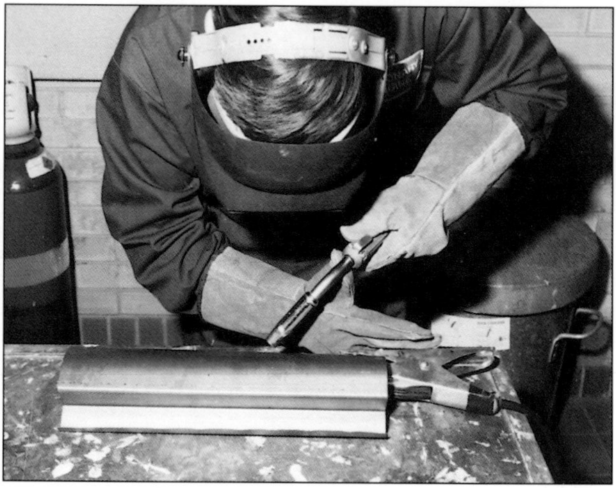

assuming you are properly protected, there is no danger. Positioning your head up above or behind the torch is hopeless because the arc is completely shielded from view by the nozzle. You must get close to the arc to see what's happening and to gauge whether or not any adjustments are required.

What you should be viewing is a pool of molten metal (spanning the joint being welded) into which the arc is maintained, carrying with it droplets of molten filler wire. This can be hard to see at first through the heavily-filtered eye protector which is why the angle of view is so important. To progress with a weld the 'pool' must be inched along the joint at just the right speed by the steady movement of the torch. Move too slowly and the entire thickness of metal will become molten and a hole will be blown clean through. On the other hand, if you race along too quickly then an insufficient thickness of the metal will become molten and a weak joint will result from a self-explanatory condition known as 'poor penetration'.

The right height

Of equal importance to its speed of movement is the angle at which the torch is held together with its distance from the surface being welded. Both factors must be kept constantly correct to maintain optimum welding performance.

The angle is important because it governs the effectiveness of the shielding gas. Remember, the MIG welding process

Correct torch angles

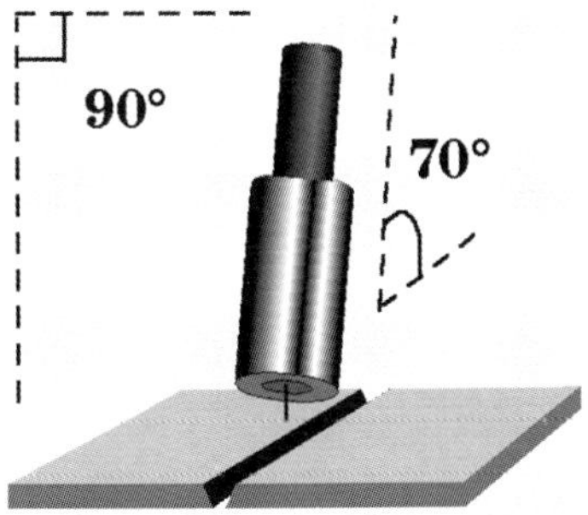

Torch making an angle of about 70° as the weld is made, but maintained vertically from side to side.

hinges on the complete protection of the weld pool from atmospheric gases. This is achieved with an 'umbrella' of inert shielding gas which is continuously maintained as the weld is made. The gas gushes from the nozzle of the torch and flows, as with the electric current, all the time that the trigger is depressed.

To ensure complete coverage from the shielding gas the torch must be inclined at a specific angle, generally accepted to be about 70 degrees (see diagram above). Operate at a shallower angle and the shielding effect will be diminished, reducing the quality and strength of the weld. Holding the torch more vertically retains the gas protection but impairs the view of the arc with all the associated problems that that will cause.

As we have already seen, the intense heat required for this type of welding is generated by an arc between the tip of the filler wire and the earthed metal being welded. Maintaining this arc, in as stable a form as possible, is obviously of fundamental importance and whether or not this happens depends upon the all-important torch-to-panel distance. A stable arc can only be maintained across a gap of a few millimetres so it's no good flailing around with the torch a centimetre above the surface because all you'll get is a lot of crackling and popping and very little else.

Have a go!

Try this for yourself on some clean, de-greased scrap material. Set the wire speed and voltage according to the instructions for your machine, switch on the gas supply and have a go. You'll very quickly discover just how fine the tolerances are! Lift the torch fractionally too high and the arc will be extinguished.

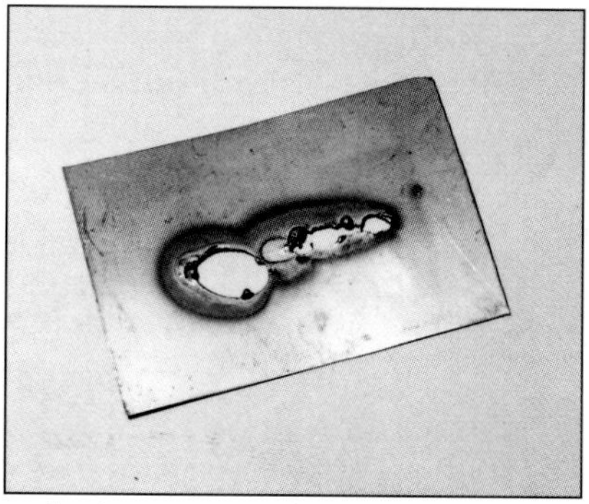

Whoops! Your first efforts will end up like this. Don't be put off, though. Keep practicing and your technique WILL develop.

Hold it too close and you risk over-heating the contact tip and damaging it. As a starting point many experienced users begin by resting the end of the nozzle on the workpiece, angling the torch to about 70 degrees and then pulling the trigger. As soon as the arc lights the torch is lifted slightly (2-3mm/1⁄16-1⁄8in) from the surface and progression down the joint begins.

Maintaining a stable arc, by keeping the torch at just the right distance from the panel, is essential for uniform penetration and consistent weld strength. It's something which you will have to practice and master before tackling any welding 'for real'. You're after a continuous crackle (like bacon frying in a pan) from the arc. Popping and spluttering indicates a problem caused, most commonly, by poor technique or incorrect settings. You must be confident of your own abilities, and of the strength of the welds you can produce, before embarking on any structural repairs which might present a hazard if the joints are sub-standard.

With practice you should also soon come to realise the crucial relevance of the torch's speed of travel. I've already touched on the consequences of moving either too fast or too slowly. Burning holes really is an obvious failing and is something which everyone does – particularly on thinner metal. Poor penetration, on the other hand, is rather more difficult to recognise with an untrained eye.

Is it strong enough?

The simplest way to diagnose poor penetration is to inspect the joint from the reverse side. If the weld is good then the underside of the bead will be clearly visible along the joint. If penetration is poor there will only be odd patches of weld showing through and some stretches of the original metal may look completely unaffected. Such a weld will be weak and certainly below the required standard for important, load-bearing joints. The problem comes, of course, when you're unable to inspect the reverse side of a weld. Welding a replacement outrigger on to a chassis box section is just such a case in point. With this type of job then it's terribly important that you know what you are doing. Confidence in the finished weld is vital where it cannot be fully inspected, and such an assurance can only be offered after much practice.

A useful pointer to watch for in terms of weld quality is what is known as the 'heat-affected zone'. This is the area of discolouration on either side of the weld bead itself. Under ideal conditions, when all settings are just about perfect, this should be around 10mm/5⁄16in wide on either side of welds made on metal that's up to 1mm/1⁄32in thick. On thicker material it will be somewhat narrower.

Which way?

One final point to bear in mind about basic technique concerns the direction in which the torch is moved. There are two established methods, 'Pushing' and 'Dragging', to choose between and the decision is primarily dependent upon the thickness of metal being welded.

The diagram below illustrates the difference between the two options. Generally speaking, 'pushing' is more suited to the welding of very thin metal because it helps reduce the risk of burning holes. The way the torch is moved means that the filler wire is being continually fed into relatively cool metal as the weld progresses. On thicker samples where good penetration is very important, use the 'dragging' technique. In this way the wire feeds into hotter metal which enhances the overall effect and ensures a good, strong weld.

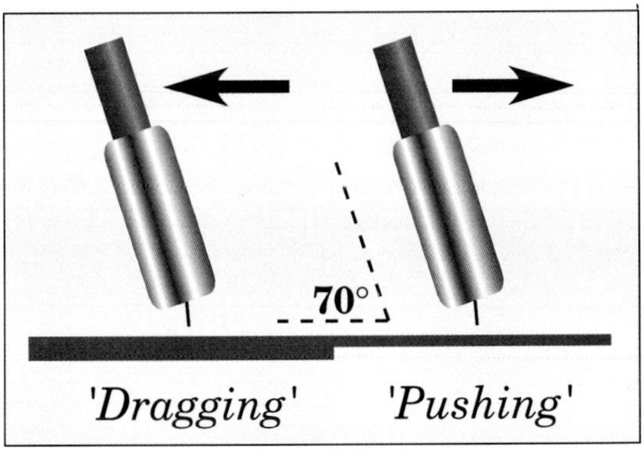

7. Beads & joints

Using a MIG welder is rather like walking a tightrope. The operator is treading a very thin line between success and failure at all times. Balance is the key. The effects of wire speed, voltage, torch movement and position must be juggled to suit the thickness of metal being welded.

Any imbalance in the system will cause a problem and being able to recognise such trouble is a very important ability which all competent operators must possess. Understanding the inter-relationship between the wire speed and the voltage setting, and how the two can be adjusted to achieve the same overall effect, is vital.

Signs of trouble

One of the most obvious indicators as to the quality of a weld is the shape of its bead. There are characteristics to watch for which provide valuable insight into the likely cause of the problem. As the diagram below shows, the ideal is to achieve a bead which is semi-circular in cross-section. However, the reality for the beginner is that, initially, you are far more likely to be producing one or other of the two alternatives!

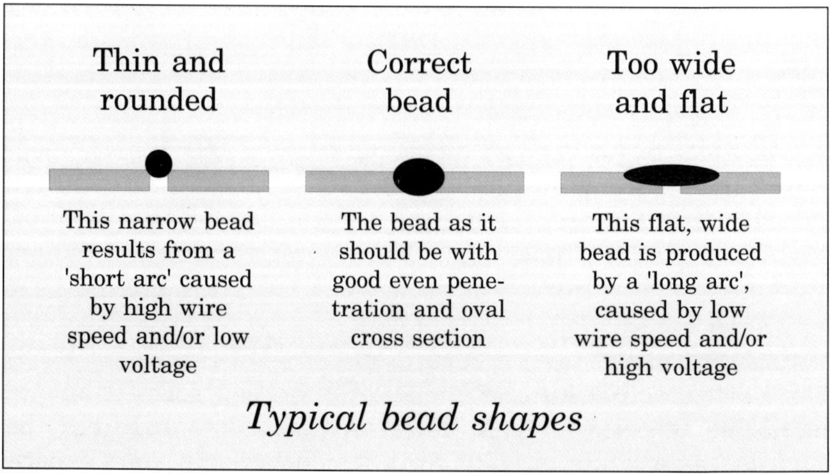

Typical bead shapes

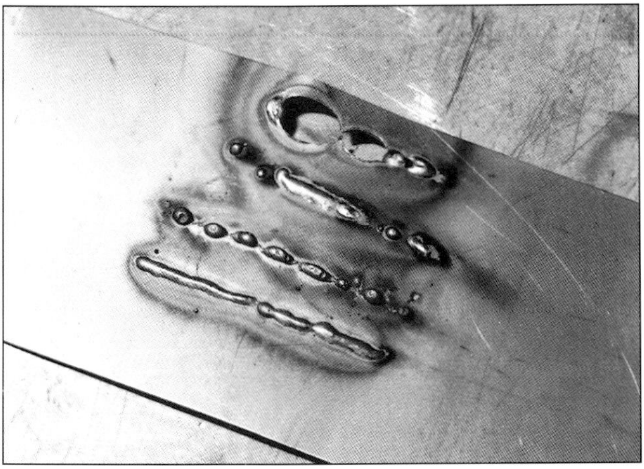

Practice makes perfect and you should certainly spend lots of time making trial runs like these before getting down to the serious stuff.

A thin, rounded bead which sits on top of the metal surface is an indication that the arc is too short, caused by too low a voltage in relation to the wire speed. The machine is simply spewing out filler wire at a quicker rate than the arc can consume it. If, on the other hand, the voltage is set too high for the wire speed, then the wire is likely to burn back up its length creating a long arc. The result of this will be a wide, low bead which will almost certainly lack penetration.

Correcting these faults is a matter of judgement which should essentially be based on the thickness of the metal being welded. Every situation will differ and so it's very hard to lay down indisputable rules in this respect. Nevertheless, perhaps the simple example which follows might help to explain the rudiments.

How does it sound?

If, while welding thin metal, you notice a lot of fizzing and popping from the arc then the voltage is set too low. The solution, however, is not simply to increase the voltage because then all you will do is blow holes. You have to approach the problem from the other direction, and stabilise the arc by increasing the wire speed.

As I mentioned earlier, most literature supplied with MIGs these days includes some information on control settings. It cannot be stressed enough, however, that this should only be regarded as a guide, a point from which to begin your experimentation. Always make some practice runs on scrap metal (of the same thickness as the metal being repaired/welded) to satisfy yourself before you start in earnest.

If no such information is provided with your machine then the best advice is to start on one of the lowest voltage settings, and with a lowish wire speed, and experiment from there. Remember that fine adjustments are far better made with the wire speed control simply because it has so much more flexibility than the voltage settings.

Getting it together

The purpose of MIG welding is to join pieces of metal permanently and strongly together. There are various types of joint used to achieve this and the choice between them depends on the application. However, before we look in detail at the options, it's worth mentioning the business of 'tacking'.

It's very important, with any form of metal welding, that the two pieces being joined remain in correct alignment as the weld progresses. To ensure this, and guard against the distortive effects of heat, it's necessary to secure them together. By far the most effective way to do this is by tack welding.

A tack is made, once the pieces are correctly aligned, by holding the MIG torch in one place and dabbing the trigger for

Tacking is an important welding technique to master. It is essential for ensuring the accuracy of a joint and for reducing the effects of heat distortion.

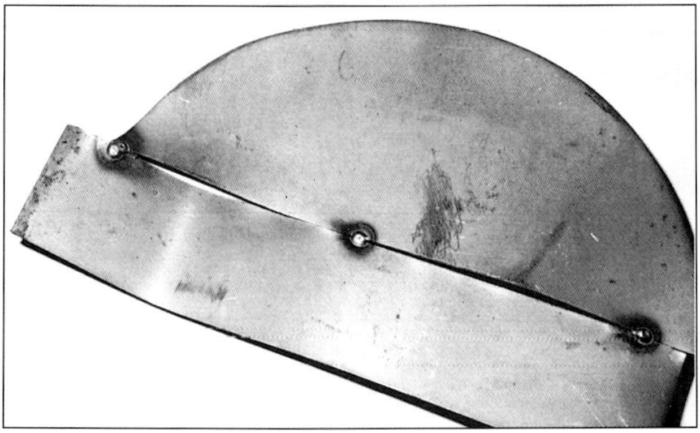

The secret of good tacking is in the spacing and this is determined by the metal thickness. Place them too far apart and this is what happens!

half a second or so. This will lay down a small weld across the joint to fasten it at that point. A succession of tacks will be needed along the length of the joint and their spacing is an important factor if they are to perform their function properly.

As a guide, metal of around 1mm/1⁄32in thickness (typically, car bodywork) will require a tack to be made every 5-6cm/1.95-2.34in. This spacing is relatively large in comparison with that recommended for gas welding simply because the heat from a MIG torch is so much more localised and, therefore, the risk of metal distortion is considerably reduced. Variation of the spacing distance will have to be made for thicker (further apart)

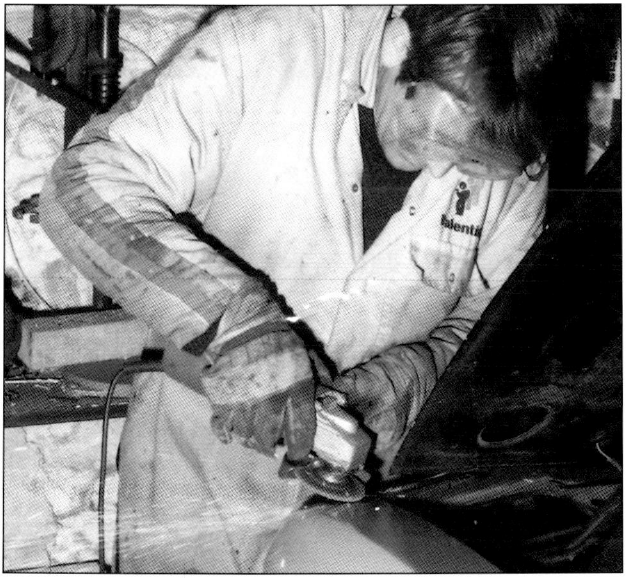

For the neatest result grind tacks off before final welding

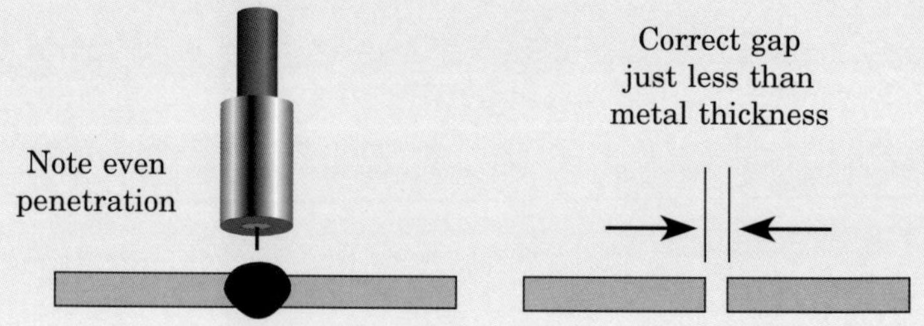

Butt joint

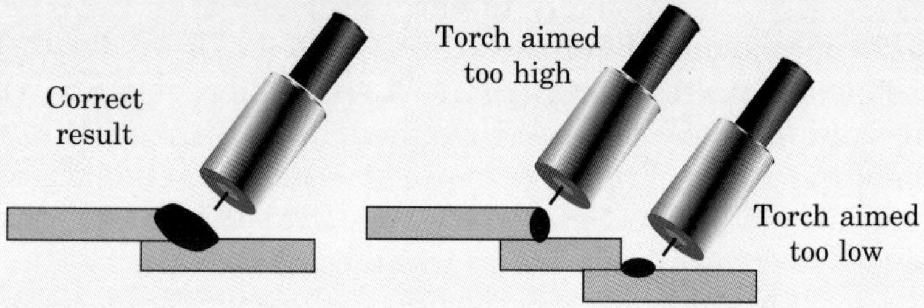

Lap joint

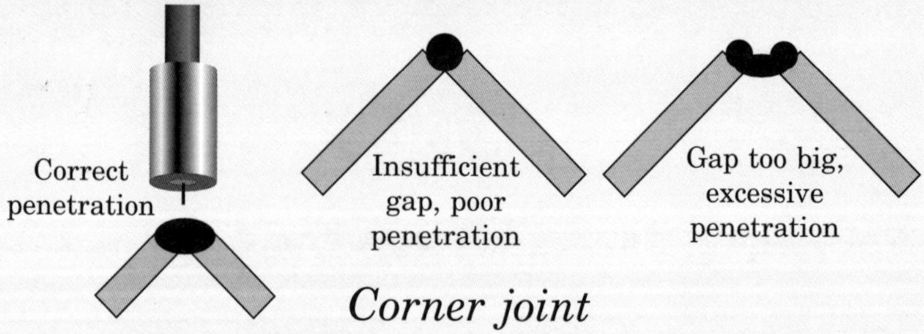

Corner joint

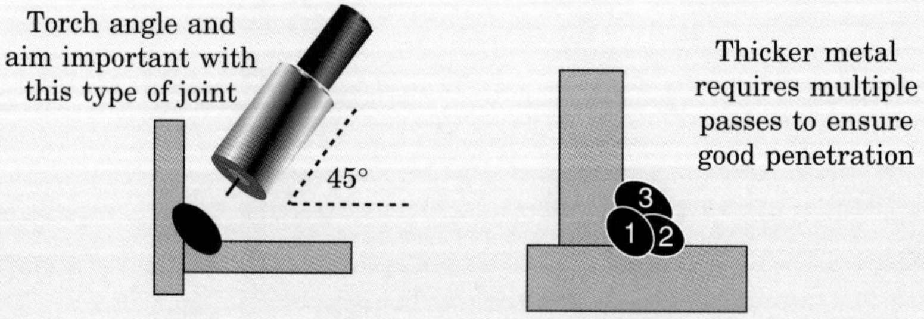

T-joint weld

and thinner (closer together) metal thicknesses.

Finally, on the subject of tacks, a useful tip is to grind the tacks off smooth before you actually start welding. Doing this will make life considerably easier when it comes to welding along the joint over the tacks, and will produce a much smoother, more presentable weld afterwards as well.

Joint account

The diagrams reproduced on page 36 opposite show the main types of welded joint found on the motor car. The most common two which you are likely to encounter are the 'butt' and the 'lap', both of which have their own peculiarities.

The butt weld is probably the hardest of the lot to master. It involves the welding together of two pieces of metal lying edge to edge, so there is no room for error. On thin material this is the type of weld which sorts the men from the boys in terms of ability. It also provides a stiff test for many smaller MIG units which are often unable to work at a sufficiently low voltage setting to avoid blowing holes.

The only way to get round a persistent hole-blowing problem, assuming that the unit cannot be turned down any lower, is to adapt your welding technique. First, you can increase the

On thicker metal (over 2mm/1⁄16in) the two edges being welded should be 'V'd' to ensure complete penetration. It ma also be necessary to make several passes with the torch as shown here.

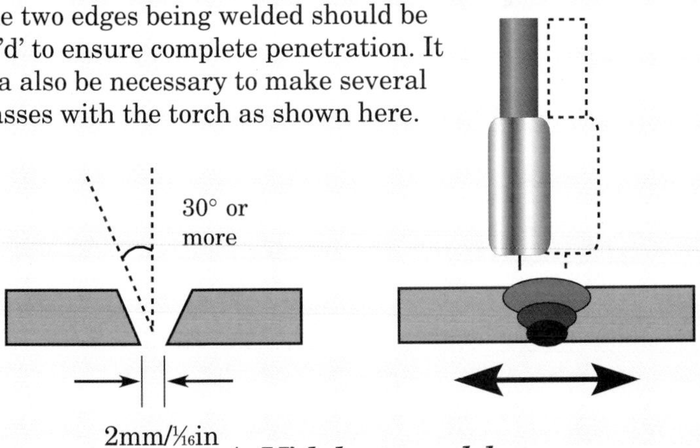

A V'd butt weld

number of tacks along the joint and also, instead of attempting the whole joint in one pass, you can employ a pulsing action by triggering the torch. Weld a centimetre or two, stop, allow the metal to cool momentarily (to a dull red), then continue on again. Obviously, the distance welded between stops will depend on the metal thickness and is something that you will have to discover for yourself by trial and error.

Another important point for consideration regarding the butt weld is the distance to be left between the two pieces being joined. On the thinnest of metal sheet there is no need to leave any gap at all but with metal that's 1mm and thicker, you must leave a gap roughly equal to the metal's thickness.

The gap is necessary to ensure complete penetration of the weld to maximise its strength. With even thicker metal (2mm/1⁄16in and above) the edges must be angled to create a 'V' between them. In such instances it's also likely, particularly with smaller MIGs, that the joint will require a number of passes before it's finished. The correct way to approach this is illustrated in the diagram on page 37. Note how the second and third runs have to weave from side to side to fill the space.

Aim is important!

The lap joint is also a pretty common occurrence within the construction of a motor vehicle. The two pieces of metal being joined are overlapped and a bead of weld is run along the right-angle created between them, as shown on page 36. Creating a good lap weld calls for accurate torch handling. The filler wire must be aimed right into the corner if the weld is to be completely effective. Running too high or too low will result in uneven penetration and a sub-standard, weak weld.

The corner joint is a variation on the butt weld theme although it's slightly easier to execute. The two pieces of metal are set at whatever angle is required and the crucial thing is that the penetration gap is of the right width – ideally just under the metal's thickness. If it is too wide then the weld will sag through, if it's too narrow then there will be insufficient penetration (refer again to diagram on page 36).

The T-joint is used to secure two pieces of metal at right-angles to each other and is made along similar lines to the lap weld. On thicker material the multiple-pass method has to be adopted to ensure good penetration and, consequently, maximum joint strength.

Plug it

Plug welding is the one other joining technique worth considering here. It's easily done with a MIG and has the advantage of re-creating the effect of the traditional factory-made spot weld once it's been cleaned up. Holes should be punched or drilled in the top piece of metal and then the two must be clamped tightly together in the correct position. The torch is held directly above each hole which is then filled with weld.

The torch should trace out a small circle during the second or so that each plug takes. This will ensure even filling and good penetration. It's likely that the wire speed and voltage will have to be turned up slightly higher than normal for this operation. Plenty of practice on scrap material (of the same thickness) will be required before this technique is mastered. Burning holes and over-filling are the obvious mistakes most commonly made by beginners. The former is a disaster, of course, but the latter can always be ground away to produce an acceptably neat result.

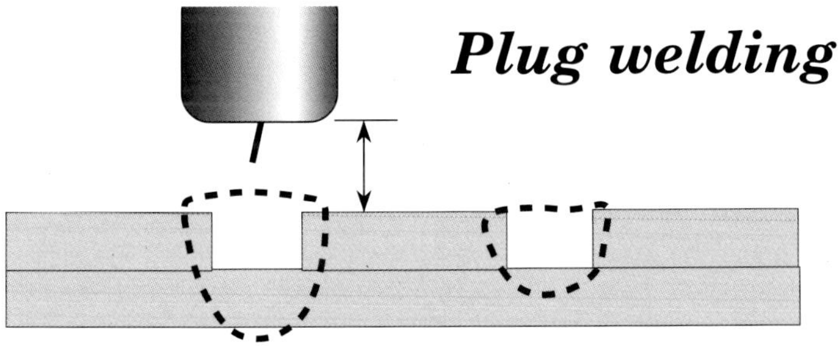

Plug welding is a very useful technique, especially when you are trying to re-create that original 'factory' look. Success will require plenty of practice on scrap metal. Torch height is the all important factor. Too far away from the panel will result in poor penetration (on right) while holding it too close will blow a hole. A good 'plug' will show penetration clearly on the underside (example on left).

8. Common faults

Mistakes are inevitable when you first start getting to grips with a MIG welder. It takes time and patience to develop an effective technique but this learning process can be speeded up greatly if you are quick to recognise where and why things are going wrong.

The ability to diagnose faults is very valuable to a welder, whatever his or her experience. Appreciating what's going wrong and knowing how to put it right really is half the battle and such knowledge only comes from a thorough understanding of the processes involved.

The idea of this chapter is to guide you through the mistakes which you are most likely to make and to offer solutions for them. Regard this information as a guide and adapt it to suit your individual circumstances and, remember, MIG welding relies upon the interplay between a number of important variables. Varying any one affects all the others so think before you act!

The hole truth!

The easiest mistake to make when learning to use a MIG welder is to burn a hole. Nine times out of ten this can be put

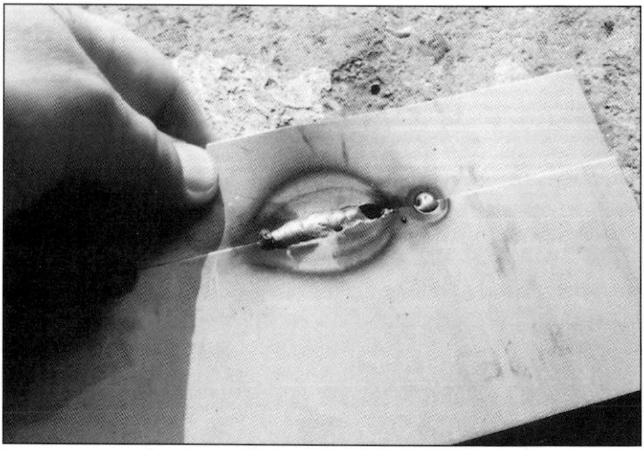

This sort of mess will be a familiar sight to all beginners! Hole blowing is impossible to avoid when you are learning to use a MIG welder so don't despair too much!

down to poor technique, with careless torch movement and control settings usually being the major culprits.

Most commonly holes are burnt because the torch is being moved too slowly. The speed needed for welding body panel-thickness metal is surprisingly fast, particularly compared to gas welding. The intensity of the heat source means that the slightest dawdle will result in a hole being blown clean through the metal and this can be infuriating.

Apart from over-slow torch movement, incorrect voltage setting is the other main contributor to holes. Selecting too high a power level is bound to cause damage which is why practicing on scrap before you start is so important. Making all the mistakes and establishing the correct settings by test-welding on metal which doesn't matter is by far the best approach.

Gently does it

It is best to start on a low voltage setting. You can always increase it but there's no going back once a hole is burnt. In some cases, if you are trying to weld particularly thin metal with a very small MIG unit, then it's likely that you won't be able to avoid burning holes. This situation calls for a change of approach and the adoption of the 'pulsing' technique that was described on page 38.

If you are working on thicker metal, which is able to resist hole-blowing, it's still possible to recognise if the voltage is set too high. For a start the weld bead produced will be wide and

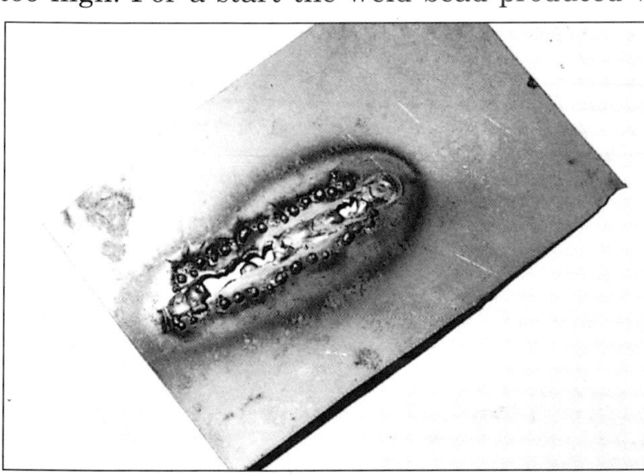

Spatter like this is a tell-tale sign of a voltage setting which is set too high.

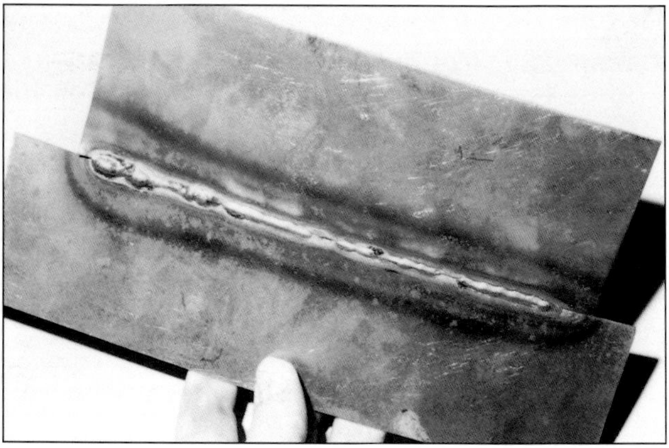

The heat-affected zone is another useful indicator as to the quality of your welding technique.

flat and around it there will probably be a good deal of 'spatter'. Spatter is best described as a peppering of the metal surface by tiny droplets of molten metal which are thrown out by the violence of the welding process. It's not so much a serious problem but more of an irritating one, particularly if final presentation is important. Spatter makes a weld look shabby and is time-consuming and irritating to grind off afterwards. Far better to avoid it altogether by getting the voltage setting right in the first place.

It's also worth noting that when working on thicker metal, the user can get tempted into moving too slowly with the torch, which has its own particular problems. For a start it will lead to a considerable build-up of weld material. Although a strong weld is likely to result, so will unnecessarily large bills for filler wire and shielding gas! Working in this way is a waste of consumables and requires a lot of 'cleaning up' time afterwards as the excess weld has to be ground away.

Another less obvious consequence of slow torch travel is an over-large heat-affected zone and the most noticeable effect of this is to greatly increases the risk of distortion (particularly on thinner material).

Insufficient strength

At the heart of a good weld should be its strength. A properly-made weld will often be stronger than the surrounding metal but a poorly-executed example can be worse than useless.

Penetration is the key to strength in a weld. Wherever possible you should make a visual check for complete penetration by inspecting the underside of the joint. On top, you see, a newly-made weld can look neat and strong even if it has not penetrated properly below – here lies the big danger.

Sometimes, however, the problem is more obvious. A thin, stringy bead is usually an indicator of poor penetration, resulting from the voltage control being set too low or the torch being moved too quickly. Other common causes are that the wire speed is set too slow in relation to the voltage, or that the torch itself is poorly directed at the joint.

The positioning of the torch, in terms of where the filler wire is being aimed, is a vital consideration that's often overlooked by the inexperienced user. After all, if the arc and filler wire are not directed to the correct place, how on earth can the weld be made effectively?

Keep it even

Remember, most small MIG units are nowadays supplied with 0.6mm/0.024in filler wire as standard. Practically speaking, this is about the same thickness as a needle, which means that its positioning has to be very precise for the best results. Unfortunately, it's all too easy to deposit most of the weld on one side of the joint or the other, leading to uneven penetration and a weak weld. This can happen whatever type of joint you are making, so be warned.

By now you should appreciate the fundamental importance of torch angle and the need to maintain a 70-degree slant in

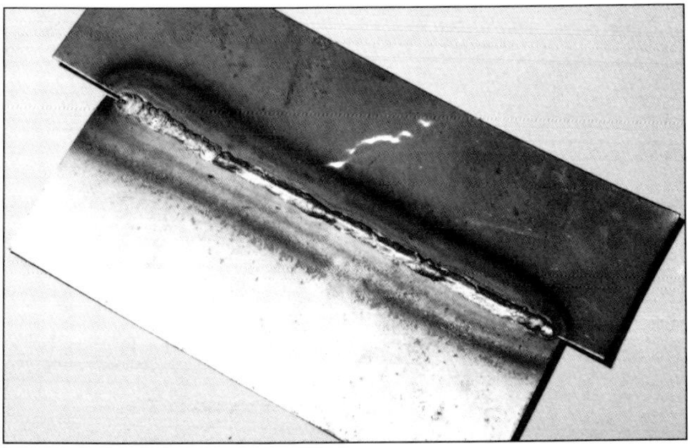

The evenness of a weld's penetration determines its strength. Inspecting the reverse side always provides the best indicator of quality.

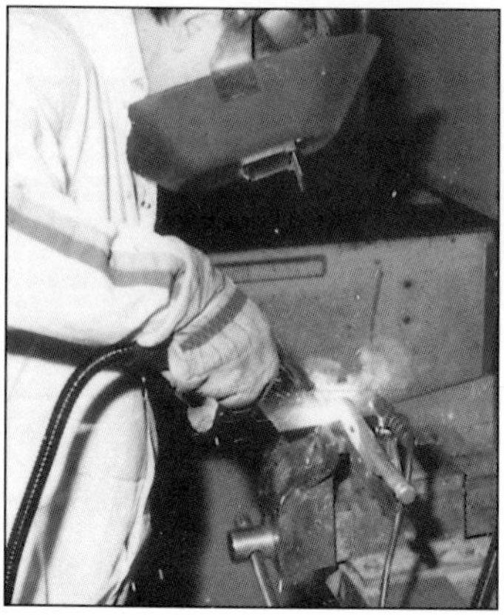

Mounting the workpiece 'comfortably' makes maintaining correct torch angles easier. Unfortunately, using a vice is not always possible.

the direction of travel. However, a further complication is that, to achieve even penetration, it is essential that the torch is also held vertically in relation to the weld (refer back to page 29). Get this right as well and the arc will act equally on both sides of the joint producing complete fusion together with unquestionable penetration.

The other significant factor which can seriously reduce the strength and effectiveness of a weld is a condition known as 'porosity'. This literally describes a weld bead that is riddled with tiny holes and is. therefore, considerably weakened.

The cause of porosity is often atmospheric contamination which points immediately to a problem with the supply of shielding gas. This defect in the weld bead is usually pretty obvious to the naked eye but, if it's not, there are some tell-tale signs to note. If the arc sounds noticeably more spluttery and crackly than usual then the chances are that you are producing

Weld porosity like this can be a real danger. Look out for it because it will seriously reduce the strength of a weld.

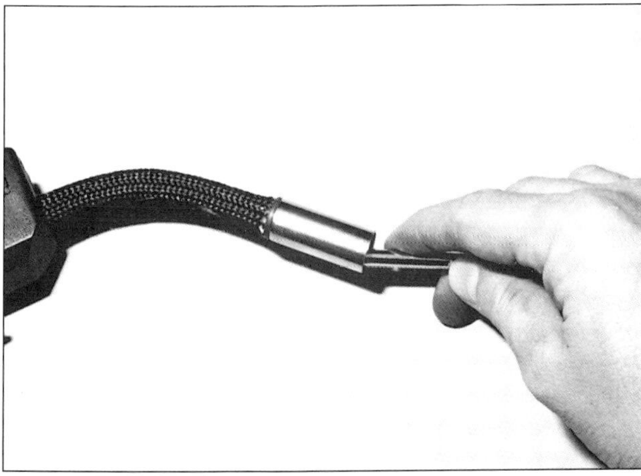

Keep the torch nozzle clean and free from spatter at all times so that the flow of shielding gas remains unhindered.

a porous weld. Excessive spatter can be another tell-tale sign of this undesirable welding fault.

Fault diagnosis

The contamination of a weld is usually caused by oxygen which comes into contact with the molten metal due to a problem with the efficiency of the gas shield. Consequently, your first move should be to check the gas cylinder contents and its valve. If these are both OK then examine the torch for any blockages within the nozzle. A build-up of spatter in this region can be sufficient to inhibit the flow of gas enough to upset the whole process. Removal is simply a matter of chipping away at it with a screwdriver or similar tool.

If, after checking the torch, the gas cylinder and the regulator, you still cannot find the problem, then another possibility might be poor surface preparation. Rust, old paint, grease and

Always make sure the area being welded is free from dirt, grease, old paint and underbody sealant etc. A grinder makes a very effective cleaning tool – wearing gloves is advisable.

other dirt left on the surface must all be cleaned off or removed with a grinder before welding begins. The final possibility, assuming none of the aforementioned is the culprit, is that it's the surroundings which are at fault. The fact that the MIG process relies on the protection of a shielding gas for its success means that wind and even slight draughts can play havoc with the results. Welding outside with a MIG unit is, therefore, not advisable under anything other than the calmest of conditions.

Don't push too hard

As you will now be aware, the filler wire is fed up to the torch by the wire drive mechanism within the main casing of the welder. A driven roller provides the motive force for this and a second roller sandwiches the wire to ensure an efficient use of that power. The tension with which these two rollers bear down on the filler wire is adjustable and the setting of this has significant consequences.

Despite the unquestionable significance of the wire tension setting, it's still something that many beginners frequently get wrong. The most common fault is to over-tension the rollers. This is relatively easy to do because the welder will work perfectly well with such a setting until, for whatever reason, the wire snags. This can be caused by an unforeseen blockage in the guide tube, by the end of the wire welding itself to the contact tip or by the arc suddenly 'dying' (because the earth clamp

Using a MIG out of doors is not usually a good idea. Even light breezes can deflect the gas shield and cause all sorts of quality problems.

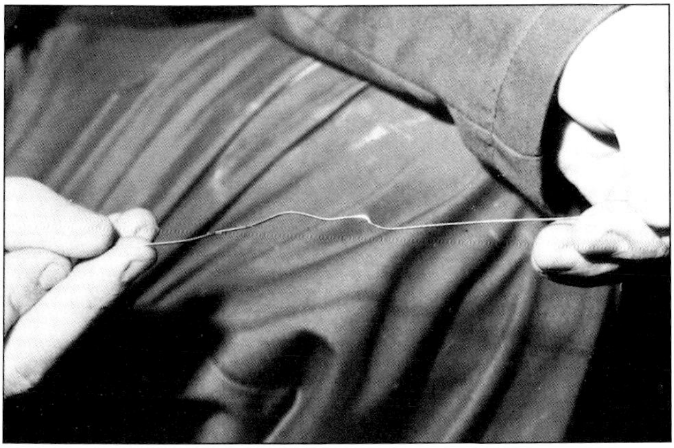

Over-tightened feed rollers can cause the wire to buckle inside the lining tube which is wasteful and potentially damaging.

has been knocked off, for example).

Whatever the cause, if the wire is suddenly stopped and the drive rollers are over-tensioned, they will continue to push it so that it quickly becomes buckled which wastes wire and can damage the torch liner. Just one kink renders that section of wire useless and the whole length, between the reel and the torch, has to be removed and scrapped which is costly in terms of both time and money.

To prevent the wire from buckling after an 'impact' with the metal surface, or as a result of a blockage, the drive rollers must be set so that they will slip if needs be (refer back to Chapter 5, page 25).

Smooth operator

A consistent and even supply of wire to the torch is essential for good MIG welding. An intermittent wire feed can be a disaster leading to an unstable arc and poor results. Dirty or badly worn drive rollers can affect the supply of wire as can an obstruction in the liner tube through which it travels on its way to the torch.

It takes surprisingly little to hinder the wire's movement. The seemingly minor problem of metal powder build-up within the torch liner can, in fact, have a serious effect on the wire feed. The other common cause of intermittent wire feed is a dirty contact tip or nozzle. Oxides and other deposits such as spatter accumulate here as a matter of course and should be prevented from building up by regular cleaning.

9. Maintenance

Like a good many machines around the DIY workshop, the MIG welder is all too frequently ignored as far as regular maintenance is concerned. The unfortunately popular 'if-it-still-works-it-must-be-OK' attitude is all too often applied to MIGs which is a great shame. You see, these units don't actually require that much attention but what they do need is important and should never be over-looked.

Torch trouble

Most problems with a modern MIG unit can be traced to the torch which is, after all, at the centre of the whole operation. Therefore, it makes sense to look after your torch. Never throw it around, tread on it or use it as a makeshift hammer. It is a precision instrument designed to carry out an exacting job on which your life could very possibly depend.

Having said this, however, MIG welding torches are generally pretty reliable pieces of equipment which normally only give trouble if they are abused or poorly maintained. The latter

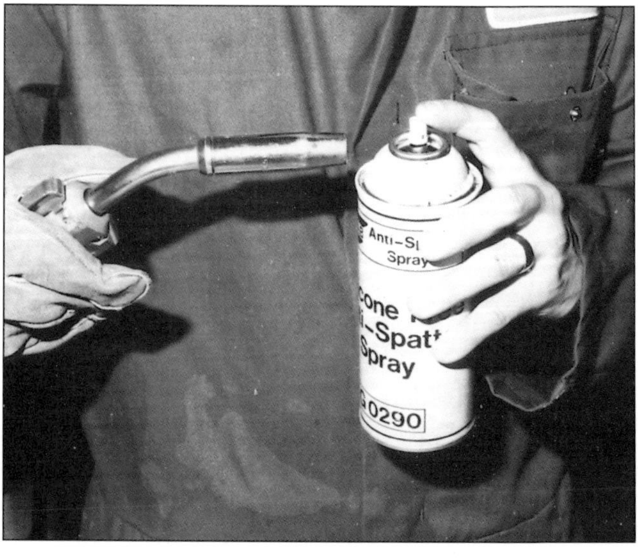

Make regular use of an anti-spatter spray on the nozzle to help keep it clear of obstructions.

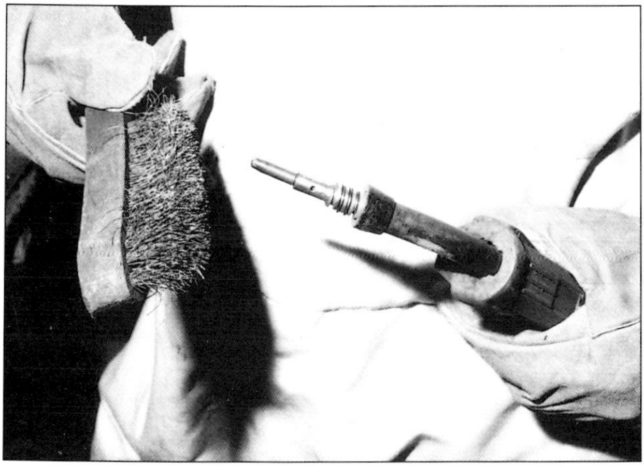

The contact tip and gas outlet should both be kept in good condition. The latter acts as a heat insulator. Check both regularly and keep them clean.

is the most common cause of failure. Look after the 'business end' of the torch and it will serve you well.

Frequently remove the nozzle and inspect the condition of the contact tip and the gas outlet. Both can be damaged or have their function impaired, by careless use or the excessive build-up of spatter and oxide deposits. A screwdriver should be sufficient to deal with most problems of this sort and the regular use of an anti-spatter spray will help minimise the problem in the first place.

If there is any wear showing on the contact tip, or it has been distorted by heat, then change it without hesitation. There is little point in carrying on with defective equipment. Remember, MIG welding is a hard enough technique to master when everything is functioning as it should – don't make life even tougher for yourself by struggling on with damaged components for the sake of money saved. The same applies to the gas outlet which doubles up as a heat insulator. Its condition is especially important.

It will be rare for a torch to need to be dismantled but if a problem does arise, what's inside shouldn't pose too much aggravation. It is all pretty basic and most repairs will be well within the grasp of the DIY enthusiast.

Catching the tube

We've already mentioned the importance of a smooth wire feed, and the effects that powder build-up within the torch liner can

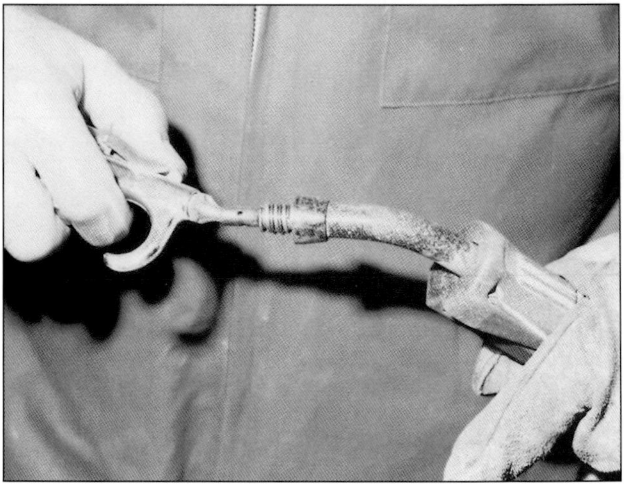

It's good working practice to flush out the lining tube with compressed air every six months or so.

have on this. There is little that can be done about the 'binding' effect of the powder within the liner apart, that is, from changing it. If you change the liner at regular intervals (every two mini-reels of wire is a sensible basis), as part of a maintenance programme, then you are unlikely to encounter such problems anyway.

Much, of course, depends upon how frequently the machine is used. A sensible intermediate measure is to 'flush out' the system with compressed air every six months or so. In this way any build-up should be cleared before it gets a grip and the life

Watch out for rusty sections of filler wire. Spools stored in damp garages can be prone to corrosion and this will render it useless where ever it occurs.

of the liner will be lengthened as well.

Your other cause for concern, in terms of routine maintenance, should be the wire spool, its mounting and the tensioning rollers. Keep a wary eye on the condition of the wire itself – kinks and rust patches are the dangers here. If you see either then, I'm afraid, the only thing to do is to cut them out. There is no point in ignoring them because they'll only cause trouble further up the line in due course.

Check also the lubrication on the wire reel's spindle. Again, this is often neglected and allowed to dry up which needlessly increases the loading on the tensioning rollers and on the motor itself. Care is needed with this, however, because over-lubrication can destroy the spool-tensioning effect of the fixing which is bad news, so be very sparing with any lubricant you apply.

Finally, you'd be well advised to keep a general eye on the overall condition of your machine. Those units fitted with wheels get dragged all round the workshop (often by the torch cable!) and tend to lead a pretty hard life. Watch for damage to the switchgear on the front panel, and for splits and cuts in the torch cable. Punctures in the cable can lead to a reduction in shielding gas pressure at the torch which, in turn, will adversely affect the quality of the weld being produced. A damaged cable should always be replaced as a matter of priority.

Cable damage, even if it's only slight, should be treated seriously. Always replace if you're in any doubt.

10. Index